MARXISM AND NATURAL SCIENCES

Y.M. Uranovsky

Marxism and Natural Sciences
by Y.M. Uranovsky

Published by Prism Key Press, 2010.
Website: www.prismkeypress.com

ISBN 1456455850

Cover by Claude Kipper

The aim of Marxism, right from its very birth, has never been the attainment of simply a philosophical perception and explanation of the existing world. It has rather been, on the basis of scientific explanation, in practice to change and upset existing relations, for the motive force of history is not the abstract criticism of ideas but revolutionary practice.

V. I. Lenin has given a profound, many-sided, concretely historical definition of Marxism, as containing" a new outlook, consistent materialism which also covers the sphere of social life, dialectic as the most many-sided and profoundest teaching on development, the theory of the class struggle and the world historical revolutionary role of the proletariat, the creator of the new, communist society". The historical and logical essence of this new world outlook and its method (i.e. Leninism-Marxism of the epoch of imperialism and the proletarian revolution) lies in the unity of theory and practice," for only thus can a really proletarian party armed with a revolutionary theory be created".

What is the connection between this practical, thoroughly revolutionary teaching, aimed at changing the world, and natural science, the science of those laws of nature which lie behind the practical activity of man when he puts the forces of nature to his own service?

The epoch in which Marx's system of views and teaching was formed was remarkable for its colossal achievements in the natural sciences and the growing social (class) function of natural science at that time.

The basis of this development of the sciences of nature was the path of conquest entered on by capitalism after the Vienna Congress of 1815. Industrial development and economic revolution gripped the whole continent. In the middle of the nineteenth century even Germany from" a moldy philistine country" was transformed into a country with a developed industry and came into the arena of world trade.

The growth of productive forces gave a powerful impulse to the progress of natural science in all its departments.

7

Let us recall how, in the small space of time between 1830 and 1848, the law of the conservation and transformation of energy was formed and given its basis in the works of Joule, Mayer, Colding, and Helmholtz. At the same time Faraday discovered electromagnetic induction. Organic chemistry developed, thanks to the work of Liebig and Wohler. In morphology Schwann's cellular theory was confirmed. In physiology we had the work of Johannes Muller and his school. In geology Lyell established the idea of evolution. Natural science passed through a period of the greatest ferment. Capitalism gave it birth and science in its turn thrust itself into the practical activity of the bourgeoisie and gave it new possibilities of industrial development. The classics of plant physiology and agro-chemistry, Senebrer, Sossior, Davy, Bassengo, Liebig, laid the theoretical foundation for rational agriculture. Thanks to the development of organic chemistry, from about the middle of the 'fifties there began a revolution in the chemistry of dyes which led to big changes in the textile industry. Pasteur, stimulated by the demands of production, made discoveries which in turn influenced both agriculture and medicine.

So long ago as the 'forties a consciousness of the importance of the social function of the natural sciences made its way into the minds of scientific workers and public persons. A discussion took place among men of science on the connection between theory and practice, science and industry. We can point to J. Liebig as to the poet of the idea of the unity of theory and practice and the adversary of the theory of "pure science".

Alexander Humboldt in the 'forties gave a verbal foundation to the development of scientific knowledge, starting from" the different degrees of pleasure induced by contemplation of nature". But this indeed is merely an "ideological aberration" which he himself exposes by recognizing the importance of the mathematical and physical sciences for the welfare of states. By his whole practical activity he rejects the theory of "contemplative bourgeois science".

8

The fact of the development of the teaching of Marx and Engels in this period of the Sturm and Drang of the natural sciences, when the achievements of science had been vastly enlarged and its social significance had grown, calls for an answer to a question of great importance for understanding the essence of Marxism

What is the relationship between Marx's ideas and natural science considered in its historical development ?

Were the theoretical roots of Marxism formed also in the soil of the natural sciences, or, on the contrary, is Marxism only a conception of history, a "science of the spirit", with which the science of nature has no inner connection ?

It is well known that it is just this latter view which is typical of the pseudo-Marxists (K. Kautsky, Max Adler, etc.). Karl Kautsky, the patriarch of the renegades, even in the years of his former "greatness" answered the question of what is meant by Marxism as follows: "I do not understand by Marxism a philosophy, but an experimental science, a special conception of society." Philosophy, the theory of knowledge, is a fine thing, "but one which has no more relation to the tasks of our Party than, for example, the vexed question of Lamarckism and Darwinism, or the question as to whether or not the atomic theory is sound".

Kautsky here advances a number of arguments. Marxism is nothing but a special "conception" of society, Marxism has no relation to philosophy, philosophy has no relation to party policy, and, finally, natural science has no relation either to Marxism, or to the policy and philosophy of the party.

Marxism is distorted in many ways in these arguments. If we turn to the last statement, then even a cursory attempt at explaining the role of natural science in the formation of Marx's ideas and the general relationship between Marxism and the natural sciences, will once more show convincingly how the pseudo-Marxists castrate the rich content of Marx's ideas, distort and contract their foundations, and so ideologically disarm the

9

proletariat in its fight for communism.

From Engels' works alone it would be possible to show the inner relationship of the different consistent parts of Marxism to the science of nature. The *Dialectic of Nature is* the most all embracing attempt at applying the method and outlook of Marx to the data of natural science. It is an attempt which is as far ahead of all that was done in this sphere by German natural philosophy and by Hegel, as the condition of productive forces and natural science in the nineteenth century surpasses the century of the French Revolution.

We shall now endeavor to analyze the problems posed here by using Marx's work and starting from his activity in the sphere which interests us.

Of course, a "vast part of the main and leading ideas in the realms of history and economics in particular" belong to Marx. But it does not follow from these words of Engels that Marx took no interest in natural science, was not equal to the development of natural science in this age, did not bring the data of this realm of knowledge within the orbit of his system of views. To-day this aspect of Marx's biography has been to some degree cleared up from the point of view of fact. He was interested in natural science while still a schoolboy, in the gymnasium at Trier, where he studied under the then famous geologist Steininger. It was the same in Berlin University where he followed the lectures in anthropology given by Heinrich Steffens, the follower of Schelling and a natural philosopher who was also an important geologist and mineralogist. Marx retained his interest in natural science to his last days. This interest manifested itself in him at various stages, in dependence on the time at his disposal for this kind of work, and in varying degree, either in acquaintance with, or study of, or active research into some scientific problem.

Marx's independent researches in higher mathematics are well known. In astronomy Marx studied Kirkwood, who discovered "a kind of law of difference in the revolutions of the

10

planets". Marx studied the relation of this law to Laplace's hypothesis and connects this discovery with the Hegelian criticism of Kepler and Newton.

In this sphere of physics Marx read Grove's *The Correlation of Physical Forces,* the work of "the most philosophical naturalist" among the English and German scientific investigators. Marx followed Tyndal's work, paying special attention to Tyndal's splitting of the sun's rays into heat rays and rays without heat.

In chemistry, particularly in agronomical chemistry, Marx had fundamental knowledge. For many years he read the literature of this subject and studied Liebig, Schönbein and others.

In biology Marx read Schleiden and Schwarn, studied Darwin critically, besides Kelliker, Trémaux, Huxley, Fraase, Helmholtz, Traube and others.

We will not here stop to examine Marx's special study of historical and experimental technology, nor dwell on the keen interest Marx showed in the conquests of applied chemistry, like the economically profitable method of obtaining oxygen invented by Rebours, or on his interest in the achievements of applied physics, such as Deprez' experiment at the Munich electrical exhibition. As for the history of science, he had a very wide knowledge, as all his works irrefutably bear witness. Is there any basis after this for speaking of the indifference or carelessness of Marx in regard to natural science ? The great thinkers who stood at history's turning-points, Bacon, Descartes, Spinoza, Kant, Hegel, generalized from the level of knowledge reached in their period and had the more permanent value, the wider the practical and theoretical basis for their conclusions, the more vividly their point of view rose above their own time. This applies even more to Marx than to any of his predecessors. Only a pygmy born to grovel in the gutters of the empyrean can look with contempt upon a giant solely because he "feels" solid ground under his feet whilst his titan's head is hidden in the

11

clouds.

What are the inner sources which nourished Marx's interest in both modern and historical natural science? What are the motives which determined this interest not as something external and accidental, not as simply a curiosity for knowledge, but as an inner necessity, so that this interest arose out of the actual general tasks of Marx's theoretical and practical activity ?

The history of Marx's concern with scientific questions may be generally divided into two periods, up to 1850 and after 1850. The essential content of the first period of Marx's theoretical activity was the finding of a basis for the materialist outlook and especially for the materialist conception of history. In the course of his work during this period he was drawn to a consideration of the problems of natural science.

It is not difficult to follow the historical course of his thought in the works collected in the Holy *Family* and in the *German Ideology.* Here Marx already advances and solves quite differently from the philosophers who had preceded him the two chief questions, what is nature-the object of natural science, and what is natural science-the science of nature.

Marx criticizes Hegel's formal, abstract, mystical conception of nature. If real nature is a natural-philosophical form of logical foundation, the reflection of the idea, then it is something lower than the idea, nature is "an imperfect being". The natural sciences from this point of view are directly bound up with theology and teleology, and can have no real importance, since they study the expression of the real creator of reality-the idea. Marx showed that the basis of this mysticism was the divorcing of nature from the practical activity of man. According to Hegel philosophical thinking must combine the practical attitude to nature with the theoretical. But with Hegel the determining basis remains the course of thought, the idea, and not practical activity, So with Hegel the picture of nature is distorted and fixed in its separation from man.

As distinct from Hegel, Marx looked at nature in its

development, in its unity with man. Man is himself a part of nature. Man is historical nature and nature is natural history. It might appear at first glance as though Marx in not yet using the category of man as a totality of social relations, completely shares the outlook of Feuerbach. In reality Marx here also, in the works collected in the Holy *Family,* had already grasped the specific link, industry, which made the foundation for new views both on nature and on its relationship to man, as well as on the specific environment which man makes for himself in the general limits of nature.

As is well known, Feuerbach also speaks of everything in nature being in a state of reciprocity, everything being relative and everything being necessary, and he sees the unity of man and nature. But with Feuerbach nature swallows man. Feuerbach does not see the historical character of the specific relationship between nature and man, the dialectic of freedom and necessity, of the absolute and relative within these relations. Feuerbach understands nature abstractly. "It follows that nature is everything, save supernatural. Feuerbach is striking, but not deep," Lenin remarks. When Marx forcibly emphasizes that "industry is the real historical relationship" between nature and man, he is laying the foundation for those views which he afterward developed with exceptional power and depth in the *German Ideology.*

We are interested in that part of these views which is related to the analysis of the reciprocity of nature and man, to the analysis of the very conception of "nature". Marx's basic thought is that nature, with the development of man and his practical activity, does not oppose man as something equal to itself and eternally unchanging. Nature develops, but after man's appearance its development is not completed abstractly outside the sphere of man and his activity, since man, whilst submitting to it, also vanquishes it. Nature is not an abstract reality with eternal "natural vocations", it is given man in historically concrete fashion through his practical activity.

This thought (or rather, these thoughts) of Marx relates

to nature taken in connection with man's practical activity, industry. For example, in the Roman Campagna there are pastures and marshland where in the days of Augustus, "one could see continuous vineyards and the villas of the Roman capitalists". This conception of nature also relates to natural science.

Neither is man connected with an absolutely unchangeable nature in his theoretical relation to nature, in natural science, which "gets its aim as well as its material, only thanks to commerce and industry, thanks to the sensual activity of man".

Natural science has to do with a relatively changeable nature ; on the one hand, as a result of the industrial activity of many generations, on the other hand (as the further development of science has shown) as a result of man's action upon it through the medium of investigation of observed processes.

The essence of the processes of nature cannot be understood without taking man's practical activity into account, which depends on the condition of productive forces and social construction. Only by starting from the practice of social life (industry, classes, social conditions) can human nature be understood as a part of nature as a whole, not only in the sense that man's psychology and ideas show their class essence, but in the sense of taking account of those natural (biological) changes to which he is subjected, when, in the process of changing reality, he also changes himself.

The method established by Marx spells the doom of naturalism in all its variations which looks on human society and man as an ordinary "child" of nature : the socio-power school (Podolinsky, Ostwald) ; the Geo-political (Rutzel, G. E. Graf, etc.) ; every kind of bio-sociological school, starting with social-Darwinism, from Karl Kautsky's attempts to supplement Marx with a doctrine of the instincts as the starting-point for the analysis of social relationships, or the efforts of the Austro-Marxists to correct Marx by the teaching of Freud, explaining religion and culture by biological factors, right down to the philosophy of modern fascism (O. Spann) which tries to base itself on a biological theory of completeness and a doctrine of races in the organic world.

Marx breaks down all kinds of teaching on freedom of will by showing that social being determines social consciousness and in this way extends the objective method to the study of the most complex social phenomena.

In place of inconsistent, abstract, materialist monism (Spinoza, French eighteenth-century materialism, Feuerbach), Marx lays the firm foundations for a materialist monism which is not abstract, but concrete, dialectical, consistent, taking account of the specific nature of human society, of all the inner connections between nature and man in their historical development. Marx gives a method and an outlook in which the dialectic of nature and the dialectic of history are indissolubly connected together.

In Marx's views the historical primacy of nature is not in any way broken. Even before the triumph of evolutionist ideas Marx establishes the following premises : the theory of creation is destroyed, as is shown by the natural sciences (geognosis) ; nature develops, it is in process of becoming even before the appearance of man ; the development of nature goes spontaneously, is immanent, self generated ; the organic world (and man) arose through generatio æquivoca ; life has not always existed as Thomson, Helmholtz and other representatives of the "absurd doctrine" of panspermy uphold. It follows that Marx understands this generatio æquivoca not as being the conception and birth of higher organisms without the intermediary of seed and parents (the medieval form of this doctrine of generatio æquivoca, spontanea aut primaria), but in the sense of self-movement, self-development, i.e. in the sense which is in accordance with the chemical theory of the origin of life and the evolutionary theory of the origin of man, established within a decade and a half by Darwin's theory.

In a deep internal connection with these new views of the object of the natural sciences, of nature, Marx develops an absolutely new outlook on the science of nature, on natural science.

Even in the works belonging to the Holy *Family* Marx analyzes, with greater power and depth than any of his predecessors (Bacon, Spinoza, the French materialists and philosophers of the age of enlightenment), the cultural-historical and social significance of natural science. Marx reproaches the philosophers for not taking into account the role and importance of the natural sciences. Natural science is not an external factor of usefulness for man or a chance factor of enlightenment. It is internally bound up with the most essential form of human activity, with practice, with industry, with the development of labor

Industry is a practical relationship of man to nature, natural science, a "theoretical relationship". Industry is the basic form of practice, natural science, the foundation of human science. Industry discloses the real powers of man, and natural science is such a "real power", "a potential of production". Marx establishes the empirical origin and practical function of natural science and apportions a very important social role to natural science.

It follows that the power of Marx's analysis, surpassing all that had hitherto been written on the importance of the natural sciences, is determined by the fact that Marx knew how to generalize with genius the objective data of the epoch. Marx did not invent theories but summed up the experience of history and modern life. He often refers to the "gifts of science" which Davy, Liebig and others made to humanity.

In the *German Ideology* Marx gives a materialist analysis of the motive forces of the progress of the natural sciences. « Pure » natural science is not a self-sufficient factor having its own history quite independent of society. Social practice has the primacy in relation to natural science, i.e.

industry, social conditions. Natural science gets from practice both its aims and the means for attaining them.

If Marx in his early works spoke of natural science as a "real power of man", then in the *German Ideology* natural science appears as a real power of the ruling class. By force of the division of labor prevailing in class society, natural science is cut away from the material process of production into an independent function, a "spiritual potential" of production. Being a factor of progress at a definite stage, this ever deepening divorce of science from industry at the same time represented the basis on which idealism penetrated the natural sciences. But the relation between science and material production is itself historical in character, being different in the age of simple co-operation, of manufacture and of large-scale industry.

Science was one of the conditions for the development of capitalism (for example, theoretical mechanics, perfected by Newton, were the condition for the development of the third period of private property since the Middle Ages, large-scale machine industry), but it is also one of the conditions for the transition to a higher social formation, to socialism and communism. Science, at a definite level of the development of material productive forces and of social development, is transformed from a condition of the enslavement of the working class into a condition for the emancipation of the proletariat and humanity as a whole.

The development of the natural sciences is not determined synonymously with the development of productive forces. If in the last resort technique and industry determine science, it nevertheless demands for its development corresponding social conditions which, in the shape of definite classes and political relations, can either assist or hold back the progress of science.

Finally, an extremely important condition of scientific progress is the theoretical premises which are provided by the work, both of all preceding generations and by that of contemporaries. Marxism consequently does not coincide with vulgar materialism in the sphere of the history of science.

On the other hand, the attempt to deduce science and its history from the social needs of this or that epoch which, in absolute opposition to Marx's views, are understood in a purely psychic sense and used as the primal starting-point in analyzing the history of the natural sciences (Gustav Eckstein, Otto Bauer, Otto Genosen, etc.), is an utter distortion of and complete renunciation of Marxism. This completely relativist theory is based on the ideas of Mach and Avenarius and is only connected with the great ideas of Marx with the aim of mocking and deceiving the working class.

What is the relation between the natural and social sciences with Marx ?

Marx's views on this question were formed on the one hand in the struggle against abstract materialism and naturalism which dissolves society into nature, and, on the other hand in the struggle against the complete divorce of history from nature (Bauer, the forefather of the Freiburg school). The divorce of the science of nature from the science of man is only possible on the basis of opposing man as subject to the objects of nature, in principle. But for Marx the monist, man, as we have seen, though a specific part, is nevertheless a part of nature. Just as nature is the basis of man, so correspondingly natural science is the basis of providing ancillary laws for the study of social phenomena. With all their qualitative difference, the science of nature and the science of man are one, for they study a single material world. They are one according to the materialist method, through applying which to the study of human society Marx discovered his conception of history.

Applying this discovery to the history of science, Marx discovered the dialectic of the history of the natural sciences. Through his analysis of the meaning of science, its social function, the motive forces of its development, its class content and the prospects of its development, Marx laid the foundation of the dialectical materialist history of the science of nature and was the first to lift the history of science on to the level of a real science.

The second period of Marx's preoccupation with scientific questions after 1850 is characterized by the fact that Marx fixes his attention on more concrete problems than those which interested him in the first period.

The wealth and variety of the scientific interests in this period of Marx's work are to be explained by the fact that Marx, on the one hand, in his work on political economy and the method of dialectical materialism, was forced to turn to natural science as a secondary science, and on the other hand by the fact that the development of science at this time was going impetuously ahead.

The attraction of science into the circle of Marx's interests proceeds by different currents.

His study of agronomic chemistry was started by his work on the study of rent. Marx in this connection, as we saw above, studied Liebig, Schönbein and everything achieved in this sphere by French authors. He followed for a number of years the dispute between the supporters of mineral and nitrate fertilizers, the struggle between the physical and chemical schools in agriculture. He was interested in everything written against Liebig's theory of the exhaustion of the soil and was acquainted with all the latest facts on this question.

Marx developed an interest in chemistry in general through his work on the method of scientific research, the theory of knowledge and the logic of dialectical materialism. This interest is inspired through the working out of the method "which lies at the basis of the Marxist criticism of political economy".

From this point of view Marx follows the revolution in chemistry and gives particular attention to the molecular theory which is connected with the names of Gerard, Kekulér and

Laurent.

Since the fundamental laws of dialectic have force in both science and history, Marx uses chemical data to confirm his methodological premises. The law of the transformation of quantity into quality which Marx examines in the transformation of the craftsman into the capitalist, he simultaneously confirms by the fact that this law is valid in natural science and in chemistry in particular, where in homological series a simple quantitative addition of elements leads to the formation of qualitatively different bodies.

It is therefore a great distortion of Marx's teaching to affirm that in the natural sciences he was a mechanistic materialist. But it is just to this that Plekhanov's attribution of Feuerbachism to Marx inevitably leads. Franz Mehring completely agreed with Plekhanov when he wrote : "Marx and Engels always remained on the philosophical viewpoint of Feuerbach, however much they may have enlarged and deepened it by extending Marxism into the sphere of history. To speak briefly and clearly, in the realm of science they were mechanistic materialists, while in the realm of history they were historical materialists."

Both historically and logically this is a very revealing distortion of Marx from the best representatives of the theoreticians of the Second International.

This distortion shows, as V. I. Lenin pointed out, a neglect of the very essence of Marxism by the theoreticians of the Second International, a neglect of materialist dialectic ; it shows a lack of understanding of the fact that historical materialism is the result of applying to the study of history the very same method used by Marx in his study of nature. It shows a superficial understanding of the deep connection between the dialectic of history and the dialectic of nature in Marx and Engels.

For Marx science served as the basis for the working out of all aspects of his method and outlook. In connection with the logic and theory of knowledge of dialectical materialism, Marx followed attentively the philosophical evolution of such a great scientific investigator as Huxley. He attended Huxley's lectures, made himself acquainted with his written and spoken work, was interested in his attitude to Comteism and analyzed Huxley's contradictory position, which approached materialism while still leaving agnostic gaps and attempting to compromise religion and science.

On the plane of philosophy and world outlook Marx was interested in the new works which showed that "the whole French school of physiologues and microscopists", led by Robin, had spoken against Pasteur, Huxley, etc., in favor of "generatio æquivoca". Marx in connection with the materialist conception of history follows science which represents the basis for his philosophical and historical views. From this aspect Marx welcomed the appearance of Darwin, whose teaching, with all its deficiencies, gave a "natural-historical basis" to his own views. Darwin gives him a new and sharp weapon with which to criticize the teachings of Malthus which are closely connected with a number of economic and political questions.

The discussion of Darwin's work is deepened during Marx's lively discussion with Engels on the work of Trémaux. In this author Marx approves of, first, the effort to determine the Darwinian chance individual changes, since with Trémaux progress arises from necessity "on the basis of the periods of development of the globe", and secondly Trémaux's effort to give a natural historical basis to such social categories as nationality by advancing the idea of the influence of the soil.

As a politician and economist Marx followed attentively to see what new productive forces were evoked by the application of science to industry (Deprez in electricity, Rebours in chemistry, Bakewell in Zootechnics, etc.). For "science was for Marx an historically motive, revolutionary force". Marx saw the inner connection of science with the concrete tasks of the political struggle and showed how the data of science, which seem at first glance to stand apart, confirmed his outlook and proved the movement of humanity towards communism.

It is, of course, hard to follow in each separate instance the motives which urged Marx to occupy himself with this or that problem of natural science. In the realm of science it is only relatively possible to isolate separate aspects or plans of Marx's interests. In reality all these aspects are mutually connected and united. One and the same sphere of science might interest Marx in different relationships. The circle of the problems which he drew into the orbit of his theoretical activities was considerably wider than the one we have sketched. Marx worked in the mathematical sphere, he was on the level of development of modern astronomy, and so on.

It is impossible to minimize the circumstance that Marx stood on the shoulders of German natural philosophy. Like Engels, he did not reject it, but critically accepted everything of value it could give him. So that in this direction also he included in his outlook the whole past development of the natural sciences.

Finally, his close scientific friendship with Karl Schorlemmer had a great importance, and particularly the collaboration with Engels who specially interested himself in working out the dialectical method in the natural sciences.

But a deep necessity penetrated the apparently accidental character of Marx's scientific studies. This was his effort to create the most all-embracing system of views, to create a consistent teaching based on the widest generalizations of theoretical and practical knowledge, as a foundation for the political struggle of the proletariat.

It is here that there grow the roots of the necessity and inner purposefulness of Marx's scientific interests, which at first glance are apparently accidental and sporadically scattered.

Once upon a time professional men of learning in criticizing Marx used to ask where in the works of Marx, "the historian and economist", his philosophy is explained, and

22

especially his "philosophy of history"? The proper answer to this question has been given in its right place. Is there any foundation for asking this question in regard to the natural sciences ? It may be said that we can find in Marx authentic statements on problems of the history of philosophy, that he has explained separate principles of the conception of nature, given an estimation of certain important scientific events of his day, but that he has no separate "philosophy of nature", that he lacks a complete Systema naturæ which answers all questions.

In such a form Marx certainly has no system of nature. Moreover, Marxism does not admit such a philosophy of nature since it puts the question of the philosophy of nature on a new basis of principle in comparison with the philosophy preceding Marxism.

Before Marx and Engels the nineteenth century had known two types of constructing a picture of the world, two types of approach to the establishment of a relationship between philosophy and science and in the very conception of the method of the natural sciences. The first type found its most complete expression in Hegel's philosophy of nature. German natural philosophy and Hegel's philosophy had the aim of uniting "the collection of evidence on final objects", which was contemporary science, of uniting this evidence on a common basis, of showing its inner connections and representing nature not as a collection of scattered forces and matters, but as a complete and organized unity.

In view of Hegel's incorrect starting-point, in view of his idealism, the task he set himself could not be solved correctly.

Hegel's philosophy of nature necessarily dissolved into "rational science", for which the empirical sciences were only the condition, but not the main picture of the world. In Hegel nature is subordinated to logic, science only regulates the course of developing conceptions. In posing the problem of the connection between philosophy and science, in making a criticism of the narrowly inductive, analytical, descriptive science of the close of the eighteenth and the beginning of the nineteenth centuries, the natural philosophers (Treviranus, Ocken, Steffens, etc.) and

Hegel played at that stage a positive part and had a fruitful influence on a number of important scientific investigators (Oersted, Schönbein, J. R. Mayer).

As experimental science developed further and the natural sciences were enriched by new data and were able to demonstrate factually the inner connections of nature, the method of the natural philosophers and of Hegel, which led to the abuse of deduction and the thrusting of artificial connections into nature, disclosed ever more clearly its own bankruptcy. After the period of the "illusory" connection of philosophy and the sciences of nature, science emancipated itself and drew apart from philosophical thinking.

The second type of constructing a picture of the world is characterized by the fact that it is applied on the basis of the empirical sciences alone, outside of all conscious connection with philosophy. Such is the vulgar materialism of Vogt and company, such is the "ordinary positivism" of Comte and of Alexander Humboldt's doctrine of the "Cosmos" which is in many ways akin to it. In dwelling for a moment on the "Cosmos" let us recall that in it the author set himself the aim of giving "a contemplation of the universe based on empiricism, on analyzed thought, i.e. on the totality of phenomena collected by science and subjected to the laws of thought, comparing and putting together these data".

This attempt of Alexander Humboldt had one positive side insofar as it expressed the necessity of comprehending the connection and unity of the data of the natural sciences, not at the dictation of an abstract idea but on the basis of actual empirical knowledge.

Owing to Humboldt's utter philosophical helplessness his "Cosmos" gave not a picture, but a mosaic of nature, not the inner connection of the data of science, but their external arrangement, not a system of knowledge, but an aggregate of observations. If Hegel's philosophy of nature was subjected to logic, then the "Cosmos" was divorced both from logic and philosophy (herein is its methodology) and hence arises its poverty in comparison with the Hegelian philosophy of nature.

If Hegel gave a method to the scientific investigator which nevertheless contained a grain of reason, Humboldt, on the other hand, disarmed the investigator into nature. The "Cosmos" was retrogressive in the philosophical sense compared with German classical philosophy, and disappeared without leaving any important traces in the history of science.

The dialectic of nature of Marx and Engels represents the overcoming of both the types of conception . of the relations between philosophy and science outlined above, types of the construction of a philosophy of nature.

According to Marx's teaching it is impossible to compose a single conception of nature and get a method for investigating nature, by starting from the activity of pure reason, for which science appears only as the condition of its movement.

On the other hand the dialectic of nature is impossible on a bare foundation of science outside of philosophical thinking. The conception and investigation of nature cannot be achieved simply by a summary of the facts of the natural sciences.

Materialist dialectic is the "total, the sum, the result of the history of the knowledge of the world". This method of investigation and understanding of objective reality in the full totality of its relations, in its development, transitions and inner contradictions, is the method which may be shortly described as "the doctrine of the unity of opposites" (Lenin).

The dialectic of nature is a method of the investigation and understanding of nature. This conception of nature is founded on the application of materialist dialectic to the data of science as they are obtained at each given historical moment. The dialectic of nature brings no artificial connections into nature and does not solve problems by substituting itself for the natural sciences. It helps in critically understanding and connecting facts already obtained, it points out the paths of further investigation and fearlessly poses uninvestigated problems.

The dialectic of nature is an organic part of the complete Marxist world outlook, the concretising of the teaching on dialectic as the science of the general laws of nature, of history and human thought. It is inseparable from the empirical sciences, on which it is based. It therefore changes its appearance with every big discovery in science. Since it is the most general conclusion from the historical development of science, the dialectic of nature 'gives empirical science a power of orientation and also directs it.

The dialectic of nature is inseparable from the dialectic of history with which it is connected by a unity of method, as two sides of a single teaching on a single, objective reality, as inseparable parts of the complete world outlook of Marx. This means that a real knowledge of nature and a conception of it as a developing whole is only possible with the knowledge of the laws and history of the development of human society which forms a specific part of nature. This means, further, that for the dialectical materialist science puts a stop to its pseudo-independent existence divorced from every aspect of social practice. The Marxian scientific investigator is consciously included in a single and inseparable complex of the theoretical and practical activity of a class which is the agent and motive force of historical progress. Science then finds its true ground and obtains a powerful impulse for its infinite development. It becomes a real weapon of struggle for changing the world and for the emancipation of the proletariat, and is transformed into a progressive and historically revolutionary force for the rapid construction of communist society.

The general foundations of the "philosophy of nature" in such a conception were laid by Marx and were systematically worked out by Engels on concrete material. Engels in this respect played a special part as one of the creators of the world outlook of the proletariat.

Marx's interest in science was not a manifestation of intellectual or scientific snobbery. The historical path of his theoretical activity has a deep logical foundation.

Materialist dialectic, that most precious theoretical weapon of Marxism, could not be the general teaching on the laws of movement in nature, history and thought, unless it had been checked by the facts of science.

Dialectic as a theory of knowledge could not have been created without the generalization of the rich experience of the history of natural science and the role of science in the knowledge of man. V. I. Lenin, that dialectician of genius, gave a special place to the history of the natural sciences (particularly to the history of the mental development of animals, the physiology of the sensual organs, etc.) in the series of other sciences "from which the theory of knowledge and dialectic must be formed". The materialist conception of history could not have been created but for the study of the laws of development of science which is a particularly important manifestation of the social superstructure. A study of the role of science is essential for the theory of scientific communism both as a condition for the emancipation of the proletariat and as a condition for the construction of communist society. Finally, the creation of political economy also calls for the study of natural science as a condition of technical and economic development, as an essential condition for the functioning of the forces of production.

The great historical and revolutionary power of the teaching of Marx, Engels, Lenin and Stalin lies in the fact that it represents the sum of a colossal generalization of all aspects of man's theoretical and practical activity, of the whole struggle of the working class. It is a united, complete, vital world outlook in which all the component parts are connected and bound together by indissolubly and incontradictably united principles.

This is precisely why it has managed to stand the test of the fire of revolution and the many-sided practice of socialist construction, both as a precious guide to action and as the theoretical foundation of the policy, strategy and tactics of the party.

The fifty years which have passed since Marx died fill an exceptional place in the history of science because of the rapid rate of progress in natural science.

Frederick Engels in his classical works discovered the inner meaning of the natural sciences in the nineteenth century, the materialist and spontaneously dialectic character of their content. So far as concerns the development of science in the last decade of the nineteenth and beginning of the twentieth century and the relations of science to Marx's ideas in the epoch of imperialism, V. I. Lenin answered this question. A whole number of bourgeois philosophers, scientists and theorizing politicians have given a reactionary solution to the question of the relation of science in modern times to philosophy and world outlook. They declare that twentieth-century science has refuted the ideas of materialism which once prevailed and which go back to French materialism of the eighteenth century, and that it has brought with it a "regeneration of the human spirit" and the triumph of idealism. For two centuries the materialist outlook has been widespread, the important German biologist Oscar Hertwig wrote, but unless all the signs of the times deceive us, we are now again at a decisive turning-point in the spiritual development of man. The two hundred years' reign of various materialist trends, against which from time to time in the past different distinguished writers have raised warning and prophetic voices, like Goethe, Fichte, Carlyle, Karl Ernst von Bar, like the physicists Fechner and Mach, is today again about to yield its place under the pressure of time to an idealist outlook.

This turning-point was announced almost simultaneously in the organic and non-organic sciences, but it was made particularly clear in modern physics.

V. I. Lenin has shown what were the conditions and causes which brought about this change and what was its true philosophical and class meaning. Twenty-five years have passed since Lenin gave his deep and all-round analysis of the crisis in science. In that time many new conquests have been made in physics, but the crisis has grown deeper yet, embracing fresh realms of science. The estimate made by Marx's great successor has not only remained unshaken but has received fresh confirmation.

In the same year as Engels died, the Württemberg professor V. K. Röntgen discovered rays which were created by the impact of electrical charges on objects in exhausted tubes. This discovery marks the beginning of dazzling successes. From 1895 to 1900 the teaching on radio-activity was created, Zeeman's effect was discovered, Planck put forward the quantum theory and thereby laid the foundations of modern physics. Rutherford established the nuclear theory of the atom and then the work of Niels Bohr began to develop the theory of atomic structure, and one after the other came a succession of pictures of the atom. In 1905 Einstein created a partial theory of relativity. In 1913 Moseley's work allowed us to penetrate further into the meaning of the connection between the elements and their arrangement in horizontal periods and vertical groups in Mendeleev's table. These works help us to understand the astonishing phenomenon of Aston's isotopes. Finally, in 1926 begins the development of wave mechanics. The impetuous movement along the path of new discoveries is not stopped, but physical thought penetrates the complex structure of the atomic nucleus. New methods of physical research lead to the reforming of the sciences near to physics, of astrophysics, chemistry, crystallography and geology.

These fresh facts and theories insistently demanded a fundamental change in all the firmly established conceptions of the old classical mechanics. Newton's mechanics were based on the conceptions of mass, energy, space and time as metaphysical substances existing separately and independently of one another.

It turned out that they are interconnected and united. Mass depends to a great degree on speed. Space and time do not exist separately, they are not forms separated from their content, matter. Impenetrability, inertia, mass, have ceased to be the unchanging properties of matter. The continuity prevailing in nineteenth-century physics has proved an inadequate and one-sided category, since the quantum theory has shown the importance of interruption in nature. The conception of the atom as the final and indivisible brick in the world edifice has collapsed, just as has the established confidence in the immutability of the elements, etc.

Failing to get beyond the old method of research and to bring forward a more perfect form of thinking corresponding to the level of scientific development in place of the old outworn form, repelled by bourgeois social relations from dialectical materialism which alone is able to replace the mechanistic materialism formerly prevailing in science, and expressing the growth of reaction "all along , the line" which is characteristic of the epoch of imperialism, the bourgeois physicists have turned to idealism and all the varieties of reactionary philosophy.

In analyzing the theoretical premises of the crisis in bourgeois science, V. I. Lenin pointed to the progress of mathematics and physics as the first cause giving birth to "physical" idealism; the second cause is "the principle of relativity, the relativity of science, a principle which, in a period of utter breakdown of old theories, imposes itself with especial force upon physicists and which, due to *ignorance of dialectics,* inevitably leads to idealism".

This argument is confirmed with especial force by modern physics.

From the relativity of the measurements of time and space fixed by modern science, physicists draw a one-sided conclusion concerning the exceptional relativity of these categories. Metaphysical reason is accustomed to a conception of the atom as an unchanging unity of mechanical structures. It calls for a stable starting-point and a final cause. But, since the atom is capable of disintegration, since the research-worker has not yet, at our present level of knowledge, been able to establish the causes of the processes which take place in the atomic nucleus, the physicists therefore draw the conclusion that it is necessary to renounce the law of the conservation of matter and energy.

From the difficulties connected with the circumstance that actual research into inter-atomic phenomena brings about changes in the object observed, a doctrine has been formed that the measurement of physical quantities in microphysics is in principle inexact and that therefore their unknowability is confirmed. As though during biological experiments, no place is found for this change in the object, which has nevertheless not prevented the penetration of the secret of, say, cariokinesis, or of the working of the muscles in biochemistry. From this well-known fact of the change in an object under investigation the idealistic conclusion has been drawn that the object has no existence at all apart from the subject (N. Bohr, P. Jordan). From historically conditioned difficulties of the methodology of physical research they draw the conclusion of "a theoretical limit" and fix absolute bounds of knowledge, as though the history of science has not completely refuted such a declaration of "Ignorabimus". The physicists, W. Heisenberg, P. Jordan, N. Bohr and others, demand a renunciation of the category of causality, though this renunciation, as Planck warns us, "is a serious thought owing to the consequences arising from it". "The new theory of knowledge", P. Jordan writes, "calls for the renunciation of all that mysticism of conceptions which was expressed as a faith in the 'compulsion', the 'necessity', in the 'comprehensibility' or the 'explainability' of natural laws and causal relationships."

The physicists, save for a few insignificant representatives of the old generation, are turning back to Kant or even more to Hume. The Machists, Franck, Reichenbach, Schlick, are utilizing these difficulties of modern physics, systematizing them and giving a basis to the reactionary conclusions of the physicists and raising them to the heights of theory.

It does not come within the task of science, in the opinion of the majority of modern bourgeois physicists, to explain processes, but only to describe them, for from this point of view the research worker in general does not know objective reality and is compelled simply to describe statistical laws of behavior

These reactionary conclusions are strengthened by class interest and are utilized as an ideological weapon of struggle against the proletariat. For example, the theorizing fascist, R. N. Coudenhove Kalergi, strives in his struggle against Marxism to work from the reactionary tendencies in modern physics and biology.

The disintegration of the atom by Hertzian rays and wave mechanics, he declares, have brought victory to idealism. Materialism is refuted. Science, from which it worked, has turned against it. It has destroyed the idol which materialism wished to set up in the place of god, the idol of matter. "With the banner of a 'scientifically' justified idealism in his hands, with God and Nietzsche on his lips, he agitates for a crusade against the Bolsheviks, those solitary allies of materialism, for a crusade organized, of course, under the leadership of an 'all-saving personality'."

In fact, any conclusions in favor of idealism and fideism are not in accordance with the content of modern physics. When a physicist deflects a-rays by an electrical or magnetic field, when he establishes that one gramme of radium discharges $3,5*10^{10}$ particles in a second, he has no doubts about the real and objective existence of rays and particles. The materialness of the world is not refuted either by the theory of relativity or by the fact that, close upon the molecule and the atom, the nucleus itself has turned out to be only a "relationship" of matter, nor by the other achievements of the modern physical sciences.

Modern physics actually confirms dialectical materialism. The theory of relativity is evidence of this in bringing us to a conception of the unity of mass and energy, of space and time. So also is the collapse of the conception of immutable qualities and elements. So also is wave mechanics which affirms the unity of interruption and continuity, etc.

Amazing as is the transformation of imponderable ether into ponderable matter, from the viewpoint of "common sense", and conversely, amazing as the absence of any other kind of mass in the electron save electromagnetic may appear to it, together with the strange discovery that mechanical laws of motion are limited to only one region of natural phenomena, while the others conform to subtler laws of electro-magnetics and so forth-yet all this for dialectical materialism is only another confirmation of its truth.

In the light of Marx's teaching the fact becomes comprehensible that, in the main, similar processes are observed in the development of both inorganic and organic sciences in the last decades. In this period in biology not only have the sciences formerly worked out been deepened, but new realms of knowledge have been discovered. To characterize the achievements of this period it is enough to recall the mechanics of development and experimental morphology, the theory of fermentations, the discovery of hormones in plants and animals, vitamins, the theory of tissue cultures and isolated organs, genetics, ecology, I. P. Pavlov's theory of conditioned reflexes, etc.

The new facts discovered in the spheres of morphology and physiology-the facts of regulation and restitution, established by the mechanics of development, the wholeness of the organism, regulated by the nervous system and inner secretions, the complexity of the processes of nourishment and motion in plants which are far from being reducible to simple laws of mechanics (the works of Max Nordhausen and Alfred Noll), etc., have called for the replacing of the insufficient, one-sided mechanical method. It was necessary to advance new principles for the connection of the growing heap of material. It was necessary to create a new "philosophy of the organic" on the basis of the factual data discovered.

In the period when capitalism had passed into the latest stage of its development, imperialism, in conditions of the growth of reaction among the bourgeoisie in its struggle against the working class and the colonial peoples, with the flourishing of reactionary trends in philosophy, science and art, the new data discovered by biology and eloquent of its factual progress, have brought about a crisis of theoretical thinking in the sciences of the organic world.

A "new course" in biology has commenced along a path sown with metaphysical and psychological conceptions, entelechy, the dominant, impulse, the super-individual soul, morphastesia, autotropism, mnema, etc. ,

A wave of reaction is rising in biology and beginning to struggle against the main biological achievement of the nineteenth century, Darwinism.

"A salutary reaction against Darwin's speculations has begun," declares O. Hertwig. "It is necessary to exclude Darwin from the series of scientific theories . . . . Darwinism has perished ingloriously," declares the Kantian Jakob von Uexküll. Eminent biologists declare that, despite the development of science, "the gap between living and non-living nature, instead of gradually closing up, has rather become deeper and wider".

In fighting one-sidedly against mechanistic methods in biology they reach the conclusion that biology does not have a method of its own, since it is heterogeneous in its logical composition and in theory yields to physics and chemistry, just as in the laboratory the biologist is gradually giving way to the engineer. It is therefore necessary to create a biology as a science sui generis, for "real biology is almost destroyed".

The ground is being prepared for the proclamation of the coming of an epoch with a new world outlook born on the biological wave (Jakob von Uexküll), for the aggression of vitalism and the appearance of a number of organically founded reactionary philosophical systems (O. Spann-the philosopher of fascism, Henri Bergson, etc.).

Vitalism (neo-vitalism) is the inevitable shadow of mechanism and its necessary complement.

On the one hand the mechanists affirm that the living is a machine, though certainly an historically-developing, complex machine; the living is an object completely dependent on external environment, its passive shadow. On the other hand, there is the opposite declaration of the vitalists that life is an autonomous subject, the laws of life are "absolutely independent and self-acting vital factors, which have the primacy over all inorganic laws; these latter must submit to the former in opposition to what has been hitherto accepted".

On the one hand, we have a violent reduction of life to physics and chemistry and the establishment from below not only of the unity, but also of the identity of nature. On the other hand is an impassable gap between the organic and inorganic worlds, or a universal teleology which establishes the idealistic identity of nature from below.

On the one hand, the mechanists state that the organism is only a sum of parts, on the other hand, the category of totality (Individuum, Totalität) is put forward, in relation to which the part is merely a subordinated means. On the one hand, causality understood one-sidedly (causa efficiens) as a renunciation of chance and expediency the reduction of consciousness to the role of epiphenomenon, a statically morphological approach to the study of organic phenomena. On the other hand, we have expediency on the basis of indeterminism (causa finalis), the introduction of psychological factors as the leading ones in the explanation of biological processes, and a one-sided physiologism, divorced from structure. The vitalists exaggerate, one-sidedly expanding certain features in the fundamentals of biology, the facets and aspects of organic phenomena, just those features which the mechanist biologists are absolutely powerless to explain.

The numerous schools created out of the break-up of biology and which are attempting to solve the dilemma, "mechanism or vitalism", the representatives of "organic biology", the Machists (Hans Winterstein), the "positive" vitalists (L. von Bertalanffy), the mnemonics of E. Bleuler, etc., are rather the smitten than the smiters, since vitalism is invulnerable from the positions of idealism or eclecticism.

Whither, for example, does the mighty condemnation of vitalism pronounced by Ph. Franck lead us ? Vitalism, he says, is only a negative concept. It is an expression of despair in physico-chemical method. "Nowhere is there a really vitalist biology. You can construct nothing out of cries of despair."

Actually his threats to vitalism are anything but terrible. In fact, as we know, science for a Machist is only the simplest description of phenomena-according to the principle of economy of thought. Science has to do with experiences and the symbols adapted to them. For Ph. Franck considers that the nucleus, protoplasm or reductional division, for example, are only relations between symbols. Why then not construct a biology as a science utilising the conception of "induction" borrowed from Uexküll or Driesch's super-personal entelechy ? Franck can say nothing at all convincing against such a possibility. Moreover, he has to recognize as theoretically possible the construction of biology out of teleological representations. The Machist cannot dispute that entelechy is a more economic symbol than the categories of scientific biology, but god or goblin is a simpler representation than Uexküll's "psychoidal law of induction" or Driesch's unrepresentable entelechy, to which we might apply Mephistopheles' words

> With thought profound take care to span
> What won't fit into the brain of man.

Trying to work from the most recent achievements in biology, vitalism tries thereby to prove that it is corroborated by the Conquests of science. But the reliance on Spemann, Jennings, Yerkes, etc., is purely verbal. The whole "philosophy of the organic" of the vitalists is reduced to the fact that the laws of the material world discovered and established by biology are connected in a purely verbal way with "psychoidal induction" and entelechy. For example, Academician I. P. Pavlov's well known teaching which permits us by using a strictly scientific method to establish certain essential laws of the functioning of the higher nervous activity, and which is not only materialist but a teaching objectively confirming the laws of dialectic, isalso, it appears, called on to confirm vitalism. "Pavlov's well known experiments", Uexküll writes, "are particularly fitted for the study of induction." 17) But the fact is, however, that this induction is anything but fitted for a weapon of biological research from the point of view of the teaching of Pavlov himself, since this induction is a metaphysically reversed and mystified conception of the reflex. Uexküll tells us concerning this mysterious induction that it is a "psychoidal law"and thus reveals that either he will not or cannot understand what are the reactionary tendencies in physiology against which Pavlov's teaching on conditioned reflexes is aimed.

Neo-vitalism seeks confirmation in the data of comparative physiology, particularly the physiology of the organs of the senses. With this comprehensible aim Johannes Müller's law of the specific energy of the sense organs is adapted in an absolutely one-sided fashion in the spirit of "physiological" idealism and raised to the rank of "the fundamentals of all biology". By bringing under it all the facts of modern physiology, including Pavlov's teaching, it is not hard to reach the conclusion of the autonomy of life and the primacy of vital factors.

These attempts by the vitalists to work-after Driesch's experiment in the sphere of the mechanics of development-from the facts of the physiology of the sense organs, show how true was Lenin's brilliant analysis which established the problem of relativism as the methodological core around which the crisis in bourgeois science revolves. In fact, for those who hold a metaphysical standpoint it is particularly difficult to grapple with the element of subjectivism and relativism which exists in the data of the sense organs. On the other hand, the data of the physiology of the sense organs which are eloquent of this relativism are the more attractive for those who strive to justify a "physiological" or any other form of idealism.

Lenin's analysis of the crisis of the physical sciences is fully applicable also to the explanation of the condition of modern biology. As in physics, the theoretical premises for reactionary inclinations were created by the very progress of biology. As in physics, in place of the mechanical method a deeper form of thinking was called for. A fundamental refashioning of the main categories of biology was demanded, of life, the individual, causality, expediency, development, form, function, etc.

The majority of research workers in biology have also, under the pressure of the social conditions of the imperialist epoch, having no knowledge of dialectics, turned towards reactionary philosophy. This turn to reaction in theoretical biology has a different expression. The ranks of the supporters of mechanical materialism have grown thinner, whilst the theoretical biologists, resurrecting the anything but advanced aspects of the teaching of the great investigators of living nature, Lamarck, K. E. von Bar, Johannes Müller, appealing to the shades of Kant, Schelling, Ocken, Mach, etc., have created many schools of different idealist shades from Machism to Driesch's metaphysical vitalism. The condition of the bourgeois philosophy of biology is largely characterized by the style of ideas which Hans Driesch is rather actively propagating. His Schillerian "Hans metaphysicus, a famous thinker, a great little man" is preaching from the roof of the vitalist tower his philosophy of the organic constructed on the data of biology plus an inconceivable entelechy which after death is transformed into a superentelechy, as is demanded by "the doctrine of immortality in its Indian form, consequently, by the doctrine of the transmutation of souls".

The crisis of modern biology is deepened still further by the fact that the crisis in the border sciences, in physics on the one hand and medicine on the other, influences biology, strengthens the chaos of conceptions and chokes it with incorrectly drawn conclusions even in the sphere from which they are transferred.

The representatives of physics have dealt a heavy blow at modern biology by attacking determinism and preaching freedom of will, which they deduce from the apparent indeterminism of infra-atomic processes (A. Sommerfeld, N. Bohr, P. Jordan, etc.).

Jordan, for example, openly considers it unreasonable, in view of the fact that we do not know the basis of the disintegration of the atom, "to ask the question of on what basis this mutation has taken place just at this time, and not thousands of years before".

The reactionary views of the physicists have given direct support to the vitalists and upset the mechanists. In illustration of this argument it is sufficient to recall the name of Ludwig Rhumbler. This famous mechanistic biologist, who for many years has labored to explain the most complicated biological phenomena as the playthings of physico-chemical forces under the control of natural selection, is now beginning to overestimate values and surrender to Hans Driesch. Taking the word of A. Sommerfeld for the fact that indeterminism is observable in the atomic system and a purposive foresight is shown by its particles, Rhumbler draws conclusions which he applies to biology. He admits that an entelechy capable of a mechanistic interpretation may be accepted. He is inclined to suppose that an entelechy is already given potentially within the atom in the shape of the energetic factor.

This slipping into the position of extreme vitalism, panvitalism, is particularly significant in a mechanist.

Just as in physics, so also in biology the latest achievements of science disclose the insufficiency and limitations of mechanical materialism, but they completely confirm dialectical materialism.

All the recent achievements of biology, the mechanics of development, the theories of ferments and vitamins, the facts of endocrinology, genetics, the theory of conditioned reflexes, etc., are a complete refutation of vitalism.

As a concrete illustration we will recall the events connected with the works of Spemann and his school.

These experiments established that the dorsal lip of the blastopore of an embryo of an amphibian when transplanted into the undifferentiated regions of other amphibia becomes possessed of the capacity of inducing a development of the nervous system, chords and mesoderms.

The nature of the action of the spheres of an embryo (Spemann's "organizational centers") was unknown until recently. The vitalists, always ready to speculate on phenomena still unstudied, hastened to declare that "Spemann and his pupils have shown in recent years the amazing multiplicity of, cases of harmonic equipotentiality".

So Spemann's organizers were to prove in this way the all powerfulness of entelechy.

But the recent works of Holtfreter, who got induction by transferring "organizers" killed by heat, frozen and dried, and the analogical works of Bautzmann, O. Mangold and Wemeyer compel us to see a chemical basis for the phenomena of independent development.

Materialism has triumphed again. A crushing blow has once more been dealt at vitalism, which is not only "a lazy", to use Claude Bernard's expression, but also a deeply reactionary conception of modern biology.

The achievements of modern biology have brought

triumph to materialism, because they explain the objective laws, the material bases, the conditions and causes of the morphological and physiological processes of a single, developing, organic world, because these achievements enlarge the theoretical basis of plant science, animal science and medicine, that is of the practical activity of man directed towards the mastery of the forces of nature.

It is precisely dialectical materialism which is confirmed by the achievements of modern biology. It is only materialist dialectic which gives a method of research, and it is the conception of unity of opposites which is the law of the processes of the organic world (assimilation and dissimulation, autonomy and correlation of organs, etc.). Materialist dialectic allows us to understand the element of relativity, the subtlety, the fluidity of the categories of biology (genus, species, individual, etc.).

Materialist dialectic is confirmed by the whole movement of biology as a science taken in its whole and compelling us to see the unity of the organic world in its inner connections and reciprocity. During the nineteenth century the two chief departments of biology, morphology and physiology, developed in deep separation from one another, to the mutual harm of both. The principal significance of the opening up of a new sphere in biology, experimental morphology, lay in the throwing of a bridge (as was seen by such a thoughtful biologist as K. A. Timiryazev) between these two completely separated spheres.

The further development of biology has still further narrowed the artificially created gap between the morphological and physiological sciences.

Endocrinology and its connected morphogenetics show the unity and reciprocity of form and function and compel an understanding of the unity and connection of the morphological and physiological sciences.

Lenin has shown that the content of dialectic must be

checked by the history of science and not by separate examples.

The history of the development of biology during the last decades furnishes convincing proof of the depth of thought of this dialectician of genius.

Thus modern natural science confirms from all sides Marx's immortal ideas. Just as the inner meaning of the achievements of science confirms the materialist dialectic of nature, so the present condition of science and its social role confirms the correctness of the Marxian conception of history.

In the countries of capitalism, where once Kepler and Galileo, Descartes and Newton, laid the foundations of modern science, this science is to-day in a state of serious crisis, accompanied in certain parts by complete stagnation and sharp decline.

The external history of this crisis and its manifestations have been fairly well described by bourgeois savants who continually return to this painful theme.

The old ideas and conceptions are utterly destroyed in the physical and biological sciences. The numerous tendencies created by the break-up of the old science attempt to advance new conceptions to unite the mass of facts discovered in the progress of science. But nothing but "chaos" (L, von Bertalanffy) and "confusion" (M. Planck) result from the search for a method.

The outlook of these scientists is distinguished by its reactionary character, its pessimism and direct connection with teleology.

The physicists, like Sir James Jeans, declare the universe is finite, and proceed to the conclusion of the existence of a mathematical creator of the universe.

The biologists support a general teleology (holism, emergent evolution). They speak of the inevitable degeneration of civilized man (D. Kotzowsky), of the mystery of the organic world (Charles Rickety, of the immortality of the soul (H.

Driesch), etc.

In place of rationalism we have intuition, in place of determinism, indeterminism, the mechanistic picture of the world has yielded to the organistic. Romanticism, mysticism, pessimism and fatalism, are growing. On the one hand, philosophical thought is going into a decline, since it is incapable of generalizing accumulated material; on the other hand, scientists are afraid of philosophy, "for philosophy is the opium of science". Positivism and Machism are growing, different schools are reviving the teaching of Berkeley, Hume, Schopenhauer and Schelling, and Nietzsche's "blond beast" is opposed to the mighty figure of Marx.

Max Planck, the physicist, denying the crisis in words, gives interesting indirect testimony of its existence. Planck indeed confirms the presence of a crisis when he is compelled, in retreating and yielding his positions, to defend the causality and objectivity of the physical world. He confirms the crisis in bourgeois science when he speaks of the confusion prevailing in science and complains that science is being overwhelmed by the activity of all sorts of fantasists. He exposes the anarchy and class character of bourgeois science when he regretfully states that these fantasists are assured of support "whilst on the other hand valuable scientific research workers with rich prospects are compelled to limit themselves or to cease work owing to lack of means".

On the other hand, the testimony of the biologist Hans Driesch is interesting. Even before the flames of the Reichstag fired by the Storm-troopers lit up the progeny of fascist medievalism, he complained bitterly that in Germany at least we are living in a time when interest in scientific knowledge is importunately and rudely pushed aside in favor of very vaguely expressed "cultural-philosophical" considerations. The desires and hopes of faith are mingled with real knowledge. This period of scientific and philosophical decadence will finish sometime. And then it will once more be recognized that the natural sciences with their strict method are the refuge of real

43

knowledge.

We will put aside the unintentional humor of the apostle of vitalism who deplores the mingling of faith and knowledge, philosophical and scientific decadence.

The important thing is to note the helplessness of the bourgeois savants which appears at the slightest attempt to analyze the causes and conditions of the crisis of bourgeois science.

On the other hand, the international character of this outstanding phenomenon of bourgeois culture, its bases and causes, is quite understandable in the light of Marxism-Leninism. This phenomenon fully confirms the philosophical and historical views of Marx and Lenin.

The general economic crisis which brings near the fatal hour of the expropriation of the expropriators, sets no great creative tasks before bourgeois science.

The upsetting and destruction of productive forces when demanded in the interests of private property, the fear of all innovation which is at the bottom of the theory of "the technical exhaustion of man", hold back the development of science or else give that development a one-sided character. Scientific workers have to justify their science by showing they are not guilty of the world crisis of capitalism (Emile Borel, etc.).

The class rule of the bourgeoisie has turned into a fetter on science. The bourgeoisie has worked science enough. It can be said without exaggeration that it develops it only so far as the interests of militarism and imperialism call for it.

The social conditions of the bourgeois world are unfavorable for the development of science. The reactionary character of the bourgeoisie which has suppressed with blood and iron the revolutionary movement of the proletariat and the colonial peoples, the reactionary nature of sections of the petit-bourgeoisie evoked by the ruin of the post-war epoch, the spirit of disillusionment, fatalism, mysticism (astrology, alchemy, magic, occultism, spiritualism, anthroposophy, etc.), spreading its poisonous color over their background, all has a fatal influence on science and colors definitely the outlook which the scientific investigator constructs on the basis of modern science.

Chauvinism, the tendency to economic isolation (autarky), the Balkanising of Europe, are all fetters on the development of science. They are obstacles to real scientific generalization and the working out of a number of scientific problems which by their nature call for co-ordination, for a frankness which excludes secrecy, and for the co-operation of nations.

The division of labor, which has developed one-sidedly in bourgeois science, creates such minute specialties that they deepen the division between the different branches of even one and the same science, and the objective basis of crisis and reaction is also strengthened. The anarchy of bourgeois social relationships does not allow the planned organization of the process of research, but private property in the instruments of research, the selection of cadres from the propertied classes, monopolizes research activity, puts wide sections of workers outside its limits and is unable to guarantee the drawing into scientific work of capable and gifted human material.

If the research worker does not by his class nature express the reactionary moods of the ruling bourgeois, the external "bidding of capital" forces him along that path.

Fascist Germany with its superstition and utilization of scientific theories in the struggle against the working class (genetics, the race theory, etc.), with its persecution of everything progressive in science, with its driving out of scientific workers who do not meet the conditions of the "third empire", is not an exception in the bourgeois world.

Fascist Germany as the rottenest link in world capitalism simply shows up more vividly and nakedly the situation of science and the scientific worker in bourgeois society.

45

In the historical sense the path of bourgeois science is completed. It has gone from Bacon of Verulam who boldly declared that "Scientia et potentia humans in idem coincident", past Oswald Spengler, the sentinel at the gate of the doomed Pompeii of bourgeois civilization preaching a fight against technique and knowledge, down to similar familiars of fascism who see salvation in "collecting all books for the bonfire".

Modern bourgeois science confirms Lenin's teaching that a crisis of method is inevitably evoked by the progress of scientific knowledge in capitalist society.

This crisis in method becomes more profoundly acute and grows into a general crisis in outlook which is accompanied by a general stagnation and decline in scientific research, as the decay of the social and economic foundations of bourgeois society spreads in the period of general crisis of capitalism.

In the fifty years which have passed since Marx's death the ideas of this giant in thought have reached out in a way unprecedented in the history of the intellectual life of nations.

What do the views of the modern theoreticians of social-fascism have in common with Marx's teaching ? What do their views on the chief problems of science have in common with Marx ?

The theoreticians of the Second International themselves cynically admit their treachery in this sphere. They themselves proclaim that they have turned from "mechanical materialism to Machism and from Darwinism to neo-Lamarckianism".

Machism is the theoretical-cognitive basis of the scientific views of most of the social-fascist theorists. Their chief arguments are that a natural law is only a convenient way of describing phenomena and that any scientific picture of the world is absolutely conditioned by social relations.

These Machian vulgarisms, which have been pitilessly exposed by Lenin, deprive science of its objectively scientific meaning.

In the physical sciences the Machian arguments inevitably bring the social-fascist theoreticians to completely sharing the lot of the reactionary-minded bourgeois physicists who preach indeterminism and idealism.

Matter, mass, is only a complex of sensations. "Neither the eternal existence of mass, nor its metaphysical uncreatability and indestructibility are established; only the constancy of the relations of acceleration observed by man is revealed."

In biology the social-fascist theoreticians stand on the extreme right wing even as a fraction inside bourgeois science. It is well known that Kautsky accepts the view of the eternity of life. In the theory of evolution Kautsky takes adaption to environment as his starting-point. So his departure from Darwin towards a neo-Lamarckianism of a psychological sort is defined. This also explains why the lesser social-fascist theorists like Gustav Eckstein and Hans Haustein refer to and support the vitalists like A. Pauly, E. Rignano, E. Hering, Semon, etc.

We know how Kautsky criticizes the fascist racial theories ; this criticism rather justifies and deepens them than exposes their scientific baselessness and reactionary character.

By refashioning Marx with neo-Lamarckianism, by biologizing historical materialism, Kautsky has disarmed himself before the fascists Günther and Lenz. He is as close to them theoretically as he is politically.

There is no dirty and reactionary source in bourgeois science from which social-democratic theorists do not draw their wisdom. The famous "freedom" of social-fascist research shows itself by each one of them in his own way, with a greater or less degree of frankness, refuting and correcting Marx. Frederick Adler refutes Marx and Engels as mechanists. Max Adler directly and Kautsky indirectly, prove Marx always to have been an idealist (for Marx, according to these distorters, always started from needs, from man's purposive activity). These theorists have their shades of opinion and partial disagreements. Finally, they very often "partake of freedom" by an eclecticism which permits them to connect the inconnectible. But the general main line of their views in science is sufficiently clear and definite. It is in the main an idealist system of views. The social-fascists stand on the right wing of modern bourgeois scientific research workers and whole-heartedly share with them the burden of ideological dispersion and decline. They are "not antipodes but twins" (Stalin).

The dictatorship of the proletariat and the Soviet system bring forward new principles in the organization of the process of scientific research. Unlike bourgeois scientific research which is partially dependent on the state but chiefly in the hands of private persons and various societies (including clerical ones), thereby excluding any possibility of planning and unity in work, socialism puts forward as its principles, instead of anarchy, a planned foundation, instead of spontaneity, social foresight, instead of one-sidedness, complexity, instead of the individualism of the competitive struggle, socialist competition and shock-work.

The philosophy of Marxism, dialectical materialism, the importance of which the mass of Soviet scientific workers recognize more and more, gives a precious weapon with which to generalize the latest facts of science, to justify theoretically the science of nature.

Science in these conditions provided by socialist society, assumes a particular power which distinguishes it in quality from bourgeois science. This is its greater activity, its greater tendency towards active interference in and changing of, those processes of nature which in the conditions of bourgeois society remain elemental and unrestrained.

The fiftieth anniversary of Marx's death almost coincides with the fifteenth year of Soviet power. The development of Soviet science in the fifteen years of its existence has fully justified Marx's views. Science in the U.S.S.R. has in this period won immense victories which have allowed it to a great extent to overtake, and in some sections even to surpass, bourgeois science. These victories have been won not only in the field of the applied sciences, but also in the field of the theoretical sciences connected with them whose generalizations rise high above the practical interests of the present day. It is enough to recall the development of Soviet physics and chemistry, the study of radio-active substances, geology and geo-chemistry, the work in genetics, the experiments with mitogenetic rays, the theory of philembryogenesis, the theory of conditioned reflexes, etc. Soviet science, from a mere appendage of European science, as it was before the November revolution, has become a strong force both within the country and in international science. These victories are the more remarkable for having been achieved in circumstances of civil war and intervention, of unceasing and desperate resistance from the remnants of the bourgeoisie defeated by the November revolution.